Looking Through Stone

Looking Through Stone
Poems about the Earth

Susan Ioannou

Your Scrivener Press

Library and Archives Canada Cataloguing in Publication

Ioannou, Susan, 1944-
    Looking through stone : poems about the earth / Susan Ioannou.

ISBN 978-1-896350-22-6

    1. Geology—Poetry. I. Title.

PS8577.A73C74 2006   C811'.54   C2006-901619-4

Book design: Laurence Steven
Cover design: Chris Evans
Cover photos: Laurence Steven
Author photo: Stefan Ioannou

Published by *Your Scrivener Press*
465 Loach's Road,
Sudbury, Ontario, Canada, P3E 2R2
info@yourscrivenerpress.com
www.yourscrivenerpress.com

We acknowledge the Canada Council for the Arts for their support of our publishing program.

Canada Council   Conseil des Arts
for the Arts   du Canada

Acknowledgements

Poems in this collection first appeared in *Event, Inscribed Volume One, Regina Weese, stonestone, TickleAce,* and *Ygdrasil.*

Elizabeth Robins' "Pleasure Mining", *Fortnightly Review,* March 1, 1902, was the basis for the found poem "The Bottom of the Sluice-Box". Matthew Hart's dramatic article "The Queen of Diamonds", *Report on Business Magazine,* November 2001, inspired the poem "Prospector".

This collection was prepared with the help of a Works in Progress Grant from the Ontario Arts Council, to whom I express my deep appreciation.

To Dr. Stefan Ioannou and Dr. Glenn Brown goes the credit for first awakening my interest in geology and mining, and to Polly Ioannou for enriching my understanding with her insights into gemstones. Virginia Storr I thank for her onomatopoeic description of beach shells, and Douglas Purdon for his vivid personal account of going deep underground. Kim Auchinachie refreshed my spirit with her unfailing enthusiasm. As ever, husband Larry gave me his patient support.

Thanks are also due to Nicole Tardif and the Laurentian University Earth Sciences Department for their generous assistance with the use of their photomicrography lab, for the 'thin sections' used on the cover: kimberlite (front cover lower half), basalt (back cover left side), and grunerite magnetite (front cover upper half and back cover right side).

To Laurence Steven—without whom—my gratitude for the insight and care with which he brought the publication of *Looking Through Stone* to such happy fruition.

For Stefan and Polly

Preface

We are all part of a grand continuum. Further up the scale, Earth is a massive version of ourselves, running on macro time. It shifts, opens, breathes fire, and flows. Conversely, any vitamin bottle reveals the astonishing number of minerals enriching our own bodies. At the subatomic level, differences dissolve. In a universe of particle-waves, a mountain is as insubstantial as a blink.

Looking Through Stone plays with these interconnections. Some of the poems open with a scientist's view of Earth—the geologist who figures in Part I. Others are born from mythology and folklore as favoured by the poet. Throughout the collection, data and story, fact and fancy interweave.

The collection approaches rocks from four different angles. The first part introduces some basic geology. Part II focuses on the seven metals of the ancients, linked to the planets and days of the week, plus platinum and uranium. Part III delves into myths and legends revealing the power of gemstones. Part IV looks at the history and technology of mining, and its social and economic impact.

Since one aim of the poems is to inform, I have tried to make geological descriptions accurate, yet colourful and simple to read. At the same time, I have probed imagination's teachings in earlier centuries. Did some, it later turned out, have a scientific basis?

Science analyses, and art creates metaphors. Ultimately, by combining the two points of view, I hope to have written poems that are both solid and beautiful.

S.I.
Toronto, 2007

Contents

Prologue

Part I - Petra

Part II - Metallica

Epilogue

About the Author

Prologue

Why

Why fondle a rock?
It shares the rugged body
where we were born.

Our passionate shift and shove
mirror a moment
erupting in fire

slowed to eons' crystalline trickle
thickening, building,
collapsing, melting.

Great ages echo and shiver,
playing themselves out
within our puny bones.

Part I
Petra

Petra

Petra: ancient Greek for rock
layered thin as shale
or spiky as bornite

a surface
cold, lead-murky
or blue-fired as aquamarine

picked from water and crevice
to hoard, to build
to dream

secure
as a cliff-carved city
or lethal as molten gold

the palm's small, marble comfort
or a boulder
hurled

Petra: breathing, magical, loved stone,
your dips and edges whisper
time's slow burn.

Geologist

He bids on the obscure: a speck
inching across kilometres of scrub
to map and pick samples out of sediments,
or cragged above evergreens, unseen
balancing a magnetometer
to listen to rocks.

He is a gambler:
under snake bellies,
between goat hooves
he trusts silver and zinc are waiting
and surfaces like scooped cream
sprout opals,
or powdered from sunshine are sulphur,
or gritty with Mediterranean blue
hold copper.

Also, he is wary: what glitters
may be the dream
—of fools.
Grade must be tested.
What grams to the tonne?
Where too angled, too deep?

A gambler bets
against absolutes.
How much should he dare
to open the Earth
to pick at her secrets
hundreds of metres down?

Exceptions

Every 27 days—sunspots—
and into dark's universe
spit solar flares
sizzling through millions of magnetic fields
until in satellites, and across Earth,
dials on delicate instruments spin.

On quieter days
most Earth rocks read
their proper positive/negative charge.
Only a few point neither north nor south
but angle the needle any odd way
as if millennia had misplaced
such specks when continental crust
in massively slow-moving plates
folded over and under itself.

For without warning,
every few thousand years
instead of Earth breathing in,
it will breathe out,
thawing the ice caps at both poles
and making magnetic fields reverse.

Thinking Big

Though hardened,
rocks go on creeping,
but through a larger time.

Their second
spans one mortal eon,
and when the two clocks collide

an earthquake cracks
or a volcano boils over
into a million widening eyes.

Magnetometrist

He listens.
The rocks hum
how minerals are drawn to each other
and particles of gold are found
as if they yearned nearby
for sulphides' love.

He listens,
and they share
their quivering magnetism:
attractions
and repulsions
moving the underworld.

Metamorphic

The Earth is never still.
Even as it crumbles
it is building,
great plates pushing
sediments up from the oceans
or sliding them under the continents.

There massive heat and stress
flatten minerals into bands
or leaf them into layers
or squeeze their particles so tight
atomic patterns rearrange
and recrystallize
limestone roughness into marble,
sandstone into quartzite,
shale to slate.

Deepest and hottest,
diamonds are hardened.
Higher, beryls and topaz cool.
Like sulphur,
without any air some form
as minerals and bacteria mingle
or, with oxygen, are reborn
like a brassy chalcopyrite
deepened to azurite blue.

Even the oldest,
like a foliated gneiss,
after remelting into magma,
hiss back up volcanic vents
and overflow as mountains
—repeating Earth's cycle.

Igneous Rock

Five kilometres under the ocean floor
deep in the upper mantle,
red, writhing magma
pushes high through denser rock
and over many thousands of years
cools into feldspar, mica, and quartz.

Or through neighbouring strata
fluids flood scalding chemistries
that over millennia mingle and harden
into more flickering minerals—
chloride, fluoride, sulphur,
silver and gold—

until, in an earthly cycle of desire,
magma rushes upward—again
to be transformed,
for no matter how solid, how old,
igneous means to set on fire,
to burn.

The Link

No rock is lifeless,
or how could it cycle
to seethe, mound, and crumble?
An invisible anima
breathes akin to our own
a million times magnified and slowed.

For what are we,
but the calcium of bones,
the brines of underground seas
awash with magnesium, potassium, iron
filling the little caverns of our mouths
with our macroscopic plankton.

Our spirit may be as divinely simple
as a magnetic charge
attracting, repelling
subatomic cations and anions
that metamorphose
flesh's terrain.

And if our chemistry is no subtler
than heat upthrusting
a mountain range,
our physics
unfixed as lava,
what does it matter?

Except that like volcanic ash
earth is what we came from
and will become again.

Haiku

lighting up night sky
a massive red aureole
hairline cracks bubble

Devitrify

Lava that cools quickly is fine-grained,
but lava that cools very quickly
letting no crystals form
hardens into a natural glass,
opening a delicate window
into Earth

but as eons creep by,
fibre by fibre from the centre
radiate more and more microlites,
tiny tubes of feldspar and quartz
filling and clouding the pane
—a cataract in Earth's eye.

Ash Flow Tuffs

So craggy, how to imagine
this hardened outcrop,
weathered grey,
once flowed like a rainbow, thick
with biotite, feldspar,
quartz, and sanidine,
and bubbling over pumice, had welded
to streaks, or flattened dark-green
to glassy *fiammé*

or before that,
hugging the surface,
unstoppable dense gas cloud
had careened
160 kilometres per hour,
700 to 1,000 hissing degrees

nuée ardente,
a glowing red avalanche
—and pyromaniac's dream.

Hammer Trail

Further and higher
along blue-streaked rock
too glittering yellow with fool's gold,
his pick hammer has clawed
and lies across a cleft.
What it could tell.

Atop earth's dromedary hump,
part of the blue-jean sky,
T-shirt one more cloud,
he perches, large white hat in hand,
and gazes across the rising desert
where tan earth greens to sage.
Beyond, the land pales into fainter camels,
dented tortoise shells, and black smudges.
A long, narrow shadow uncoils between.

This is old-prospector terrain
that swallows rotted boards
and rusty wheels,
soundless as dead coyotes
deep down abandoned mines,
still as the endless, blurred scrub
where a red four-by-four winds
farther and farther, a speck
disappearing.

Sedimentary Rock

Near Earth's surface
whatever the sun
heats and cools,
swells, softens, and shrinks,
is dried out, weakened,
and splits off.

Whatever water and ice pick at
and winds have dropped
weathers into layers,
loose compost, clay, and sand
that grain by grain dissolve
and seep down

and melting into groundwater, form
minerals lustrous as copper
in enrichment zones,
as myriad
as there are organisms
and oxygen in ample supply,
or carbonic acid
for microbes to decay,
as motley as calcium swirled in shells
or skeletons pressed into silica
before sinking back to stone
—lithified.

Rock Sample

As he tilts the rock sample across his palm
chipped turquoise becomes
the Caribbean
glowing against the night, its porous stone
solidified waves, moon hidden behind
the dark, breathless
as that barely perceptible ache
his rational mind would deny
—*Merely chemical formulae.*

Only more years will wear away
even the hardest logic, to trust:
turquoise sings in the brain,
fissures open tough fingers to love,
and the tiniest particle hovers
always beyond explaining.

Only with age,
will analysis soften
simply to murmur *how beautiful*
and accept not fully knowing
—that deeper wisdom
of wonder.

Almost Like Love

Microscopic throughout the Earth,
between dissolving and forming
watery ions are ever drifting.

A few slip closer together,
link as a chemical lattice
and form a crystal seed.

Threading out delicate dendrites
more and more may fasten:
to multiply is to thrive.

But if too slow? They slide
back into watery waiting.
Crystal never dies.

Pegmatites

Just enough water,
and seething up from Earth's core
magma cools before it reaches the crust,
not into little green beryls or white spodumene
but into elegant tubular crystals
10, even 15 metres long.

Too much fluid, the overflow
simmers below in small cavities,
dissolving, reforming, dissolving each crystal.

Too hot, and the vapour explodes
each facet into shards so small
few fingers would ever hoard.

Proportion—just as crucial
for even a crystal
to thrive.

Geode

Even a cavity in rock
if watered with crystals and minerals
can blossom into a geode—a metre wide.

Its surface may swell
bulbous, nodular, or warty
or droop stalactitic, dripstone long.
Some may uncurl rosettes
or twist to frost-fine lace
or hardened moss.
If bold, it tufts, if subtler, tarnishes.
Its shell may ooze or crust,
even crumble like earth.

Cracked open,
sparry inside will glisten
with many smooth, little cutting faces.
Or the centre may be gritty,
unless packed solid
to the naked eye.
If oolitic, its eggstones nest
prickly as pinheads
or plump as peas.
It may stand up in columns
or mat itself with fibres
or wheel out radii.
Perhaps its innards are scaly or leafed
or breathe through mushroom gills
as lamellar.

A geode is Earth's poem, unspoken,
thick with her possibilities
words are itchy to hold.

Two Facets

How corporate, the druse.
Inside a cavity in vein or rock
its encrustations may hold water
but cling,
their minerals one
with the parent walls.

Much freer, the entrepreneurial geode:
on limestone or volcanic hollows
its crystals sprout, distinct,
their quartz or calcite
angling inward, yet
able to break away.

Mineralogy Lesson

Staring at crystals
this man is a scientist,
and when I murmur of mysteries

"Physics," he grunts.
"Temperature and speed.
That's why a crystal grows, or doesn't."

What a wonder it grows at all!
How hidden under the earth
water, light, and air gather
dendrite on dendrite, facet by facet,
becoming a geometrical bloom,
or soften and redissolve.

He smirks.
"It's ions.
Diffusion."

But I believe:
the inanimate is prickly with soul,
or why do its ions attract, heat rising,
and fasten on others
faster and faster?
Isn't that love?

"It's physics,"
he insists. "Physics."
—unaware he blushes.

Even the Words

Fingered close to the ear,
even the words are solid
enough to pan for sound.
Taste their grit and sheen.

Crustal and *grossular* scrape.
Scapolite juts its corners.
Like a spade swung against metal
klint shivers the air.

How sinewy, *plagioclase*,
while *rodingites* smartly line up,
but *diopside* droops, slides over,
an oyster slurped from the shell.

If *tourmaline* cools the tongue,
skarn is half skunk, half onion.
Be careful not to breathe too deep
or choke on a cloud of *vug*.

Precise scientific terms?
Beyond tight definition
their syllables bristle and throb
more than full of themselves.

Illusion

Whenever fingers fold around a rock,
mind may whisper *jagged* or *light*,
but gut cries *solid!*
Faith is built on rock;
riches also, even the dream
that hurtles a seeker into madness.

But let our microscopic eye
refocus from chunk, to chip, to grain,
within a ten millionth of one millimetre
and further, 100,000 times smaller
(from SkyDome scale to pinprick)
enter the atom's eerie cosmos.

Its blinding nucleus buzzes
with neutrons orbiting protons
gluoned of "up"/"down" quarks spinning
to keep from flicking into some other form
and steadied by shooting back alpha,
beta, and gamma rays beyond
the whizzing outer electron cloud.
What an atomic frenzy—Please!
Is that what *solid* means?

Yes. Too, in a given moment,
each particle both exists and does not:
at once being equally a speck and a wave.
Remember the Forms that Plato idealized?
Physicists trace them in antimatter
thrown off as atoms decay.

Look closer at the granite
chunk scratching your skin
—a fistful of electromagnetism.

He Sits

Overlooking 2,000 metres of air
and pine fuzzing down
to a blue thread,
high on a sharp white rock
he sits atop his world,
face full of light.

This man
who walks upon mountains,
who listens for minerals' hum
knows in fissures and dust there is meaning
beyond the mere metallic
excuse for his coming.

Here he can listen to light,
here he can touch light,
here he can taste light.
Here he becomes light
welded to Earth's
upturned core.

Part II
Metallica

From the Beginning

Even in prehistory,
minerals were known.
Copper and tin, gold and silver,
malachite, turquoise, lapis lazuli,
and in Mesopotamia, agate and opal
were fashioned into shields, necklaces, tools.

The earliest were named
in honour of divine powers:
gold for dawn goddess Aurora,
copper for the isle of Cyprus
where upon a giant scallop shell
Aphrodite foamed from the waves.

Much later, Aristotle's *Physics*
labelled minerals after places,
persons, and properties
—for instance, topaz,
from the Red Sea island Topazos,
plucked for the diadem of a Theban Queen,
or graphite, from the Greek "to write",
or magnetite young Magnes stuck to
while on Mount Ida tending sheep.

Their mystic light
so charmed the Middle Ages
that gems were noted and codified
by lapidaries mining stories for cures:
a stone to heal an insect sting
or lure a hesitant love

until the mid-sixteenth century
when Dr. Georgius Agricola's
De Re Metallica redefined

each mineral instead by origin,
tint, toughness, rub, and use
—a grittier truth.

Three hundred years after,
in turn, the microscope's eye
chemically classified from I to IX
Elements . . . Phosphates . . . Organic Compounds,
A.G. Werner's mineral system
students still memorize.

The Metal of Heaven
Iron (Fe)

I

Not only in the ancient Greek underworld
hammered by smith-god Hephaestus,
iron is found everywhere:
in banded ore
and in natural springs,
in fish, liver, eggs, and vegetables,
and thus by eating, spurring enzymes
and bubbling oxygen through blood,
our body a microcosm of Earth.

For Earth, too, formed
from heating and pounding
when across the early cosmos
hurtling chunks from burnt-out stars
smashed together and melded,
while gravity spiralled in
more and more iron
with cobalt and nickel,
snowballing immense protoplanets.

Radioactive still, a-sizzle
within, one liquified and spun
into our own pole-flattened sphere.
Iron and nickel sank to the centre,
welding a solid magnetic core.
Lighter, silicon rose
like hot scum to the surface
and during millions of years
as more and more magma seethed
up through a hardening lithosphere,
stiffened to thin basaltic crust.

In time, thickened, it cracked
and inched apart as floating tectonic plates
over and over reshaping themselves
into new continents.

II

Throughout Earth's eons
meteoric iron kept hurtling
250,000 kilometres per hour.
Roaring lightning-bright out of clouds,
more than 2,000 thunderstones a year
flamed down "metal of heaven".
These small blackened gifts of the gods
Egyptians and Babylonians gathered
as limonite, magnetite, hematite,
their nickel-laced iron revered
for its dark-satiny beauty in beads
and celestial strength in blades and tools.

Not all iron streaked from the sky.
In Greenland, around 2000 B.C.
bogs were oozing out nodules
volcanic basalts had heaved and shoved
upwards for the picking, believed
to be gobs of petrified blood.

By 1400 B.C., the Hittites
unlocked from Black Sea sands
the secret for washing out iron-rich grains.
Stone furnaces smelted the magnetite
to much sharper daggers and swords
so envied and in demand
that later Assyrians valued such iron
40 times more than silver
and 400 more than tin.

Iron later shod horses, hinged doors,
reinforced stone walls, bolted beams,
ploughed, and shackled killers and thieves.
By 1350 A.D.
English King Edward the Third
guarded his cook's utensils like gems.

Nowadays, smelted and curled
into graceful wrought iron
for fine railings and gates
or into a harder but brittle pig iron
thickened to manhole covers
and engine blocks,
from fungicide, inks, photography, dyes,
iron fulfills 20 times more uses
than other metals combined.

The Lunar Metal
Silver (Ag)

Imagine, the moon's tears
falling to Earth as silver,
or so the Incas believed.
Most lustrous of metals,
the purer that silver is,
the darker its tarnish,
as if it waxed and waned.
Even the blackest ring or bangle
lightened if your blood was untainted.

Recorded in Genesis,
by the fourth century B.C.
Sumerians were mining silver at Ur,
while under the Aegean Sea,
Cyclops hammered silver arrows
for hunter, moon goddess Artemis.
The Romans labelled it *argentum*,
so rare that it was bent and stretched
to fashion sacred objects and ornaments,
but also bagfuls jingled off debt.
From fourteenth century to nineteenth
silver weighed the standard for coin
—92.55% pure,
among all polished metals,
its sterling the most brilliant.

For several hundred years,
a silver vessel or sunken coin
freshened larder food and drink,
and from feasting on silver plates,
aristocratic Blue Bloods' skin
turned its ashen hue.

In sprays and gargles still it soothes,
and swallowed, calms stomach ulcers.
Dropping its nitrate into newborns' eyes
is sure to ward off blindness,
while babies suck away germs
with silver pacifiers and spoons.

Folklore?
No, hard science.
Dissolved beneath the skin,
at submicroscopic size
silver's magnetism lures
virus, fungus, single-celled bacterium,
and in 6 minutes suffocates
any among 650 strains,
even gnaws on cancer.
Why, on Earth, as in outer space
silver filters cleanse our water,
—yet its salts are poison.

Mined with copper and gold,
few metals can usurp its place
in batteries, bearings, and catalysts,
switches, membranes, mirrors, breakers,
dental fillings, and solar heating,
photography, or X-rays.

Though easy to pull and shape,
even in razor cold or heat
little eats silver away.
Tough, this metal is sensitive,
so brilliantly it can reflect
the thinnest light sliver.

Smoothing a joint's motion,
adorning a woman's skin,

or as a quick conductor
for channelling moonlight into intuition,
silver is indeed a lunar metal
—of illustrious phases.

The Power Metal
Copper (Cu)

Swallow an oyster,
and its copper traces
will make your skin bloom.
Or drink from a copper pipe,
munch seeds, chickpeas, or Cheerios
to marry copper with proteins and form
catalytic enzymes
that ferment our energy
—indeed, that keep us alive.

The oldest mined mineral,
so mirror bright the ancients
placed copper beneath the heads of the dead,
it shone in 4000 B.C.
as the first sacred chalice
smiths fashioned for Aphrodite,
her island of Cyprus inspiring the name
for copper mines so deep and rich
they supplied the Mediterranean.

Round our Great Lakes, 6,000 years ago,
First Nations held thousands of tonnes,
nuggets as fat as 100 kilograms
quarried with only stone hammers
out of 5,000 excavations
for ornaments and tools.

Yet as little as air's prolonged kiss
rubs copper's metallic lustre
from red-brown to teal,
why over the centuries, healers chose it
to balance the body's energies
or purify the soul,

and pragmatists,
to rinse out their mouths
or cleanse farm-water for fish.

Outside our skin,
it wipes bacteria from wounds,
burns out epilepsy and convulsions.
Soft among metals,
its bracelet shrinks arthritic swelling
while bending under a woman's beauty,
yet in diesel locomotive motors
hydro stations, or transformers
copper is the queen of power.

Even its slenderest wire is best
for speeding electricity and heat,
and though it so delicately connects us
copper is tough
—enough to contain
a Yucca Mountain of radioactive waste.

No wonder many industries now wring hands:
the world's pure copper reserves are gone,
never to be replaced.

The Noblest Metal
Gold (Au)

Fancy, "the sweat of the sun"
19 times more dense than water
glittering down toward Earth as gold
so precious, the Incas
measured in carob seeds,
each carat one-fifth of a gram.

So, too, in ancients' minds
gold dawned as *Aurora*
(Au its chemical sign)
while Asian ritual laid gold leaves
like skin over wood and stone
to gild a statue to life.

Set around topaz,
it scattered enchantments,
with rubies, drew energy from the sun.
A talisman of power and wealth,
to Wiccans gold held the Fire
that negativity fizzled in.

For centuries, gold
was crowned "noblest of metals"
neither water nor air could corrode,
only *aqua regia*
fuming its yellow acids
in secret for alchemists and kings.
Later ages blued gold with iron,
tinted it green or pink with copper,
or whitened it, blending nickel and zinc.

Today, as no alloys
can make it magnetic,

gold is uplifted as "pure",
yet sealed, its clear solution
in time—for no reason known—
will bristle metallic particles not its own.

So sturdy gold's properties,
draw it to a fine wire
80 kilometres long
or beat it thinned into leaf,
it will not boil, not melt
below 1,000 degrees.

Across three continents
deep in Precambrian lodes it glows,
or mingled with gravel rinsed into streams
as placer deposits in bedrock hollows,
or traces scattered in massives
with other metals.
Such labour: prospecting, staking, surveying
for hard-rock or open-pit mining,
to crush, smelt, and mint
as heirloom coins and bullion bars
vaulted right back underground
as government reserves.

Gold's litany begins in Witwatersrand,
Kalgoorlie, or South Dakota
and ends:
plated, painted, even injected
into arthritic joints or a muscle,
beaded on mirrors, encircling a vase,
or against cancer,
from radioactive isotopes
dawning again—as gamma rays.

The Liquid Metal
Mercury (Hg)

So dense it will float a cannon ball,
the Romans' liquid silver
hydra argyrum
is 14 times heavier than water
—even before it freezes
at −38.87 degrees.
But drop on the surface a chip of gold
and watch how it dissolves
like a sugar cube in tea.

Yet mercury is neither wet, nor flowing.
When tilted and poured, it rolls
glowing into beads
that won't soak through the thinnest fabric
but wobble like greasy raindrops
off a duck's wing.

Named for the Roman planet nearest the sun,
in Earth's crust, as an element
mercury is rare,
although as early as 2000 B.C.
found among Chinese and Hindus,
and later, left tubed in Egyptian tombs.

Poor ancient Greeks
who blended it in ointments,
and Romans, unwittingly, in rouge,
or Peter Paul Rubens' "Three Graces"
whose seventeenth-century cheeks
wore mercury's vermillion.
Little wiser, Plains Indians
who drew its red across the cliffs
and war-streaked braves—with toxins.

Yet alchemists framed it a mystic sign,
believing that mercury meant both
death and life, heaven and earth,
for dropped with silver nitrate in a flask,
up the glass it crusted and branched
a wondrous *Tree of Diana*.

Mined today from red cinnabar,
mercury is almost harmless
—when pure.
Not so, seeped into soil and streams
where pesticides and fungicides
lay up noxious levels
absorbed by touch or eating,
or simply by breathing in
the greening air.

Safer, its gleaming locked into mirrors,
thermometers, batteries, contacts,
neon, and ultraviolet lamps.
Kinder, as the winged-heeled messenger
speeding a serpent-wound caduceus
to heal Olympian gods.

The Oldest Metal
Lead (Pb)

Beneath the earth
in heavy, silver-laced stones
or glowing in cubed brittle galena,
its bluish-white is too lustrous
for the thud of its name
—*lead*.

Too beautiful, this oldest metal
linked to Saturn, time, and death,
to weight fishermen's nets in Exodus.
Too vibrant its Latin *plumbum*
for piping through Imperial Rome
water it sickened, to seal a sarcophagus.

Lead is the lovely alloy in pewter,
flint glass, and glistening lenses,
and pings in fragile crystal.
Yet it is common enough to brim
every lead-acid car battery
—a full 10 kilograms.

So hardy it can wrap cable,
shield reactors and X-rays, and stop rust,
lead primes and solders, or simply absorbs sound.
Beware, for lead can be lethal as ammunition,
poison babies chewing old cribs,
and mass-murder bugs.

When swallowed or just breathed in,
it frays the central nervous system
and whitens the blood,
eating away at cells as cancer
or mutating a fetus
into a monster.

—What better reason for alchemists'
centuries of futile spells
to fire it into gold?

The Vital Metal
Tin (Sn)

When bent,
tin cries out
as its crystals break.
When warmed
above 13.2 degrees,
crystals turn tetragonal and pale.

If chilled,
its silvery whiteness
crumbles to so grey a dust,
warts on cathedral organ pipes
were whispered Devil's work
or a "tin plague".

Brutal as its origins
in twisted cassiterite veins
or ground through river beds and ocean floors,
in biocides tin kills
undersea bacteria and fungus
—and anything it touches along shore.
How fitting, Celtic craftsmen long ago
from secret mines in the Scilly Isles
fashioned it to knives and swords.

Yet lightweight,
harmless in its whiter nature,
tin coats cookie sheets and grocery cans,
solders drinking-water pipes,
or as fluoride toothpaste
shines up fillings.

In compounds, tin suppresses smoke,
yet lubricates bearings and toughens

molten glass in double panes.
Beautiful and soft,
for artists it alloys as pewter
or bronze in ornaments, and bells
vibrant as the names
Saxon *tin* and Roman *stannum*
humming from a miner's tongue.

Even at its coolest, it attracts.
Blended with niobium in wire,
a single kilogram of tin fans out
fields of force more gripping than
a 100-tonne electromagnet
—apt talisman for wealth.

For centuries, in secret, alchemists
besides distilling dross for gold
invoked tin's mystic sign,
their highest goal, to flame eternal life
for, in firing up their crucibles,
they'd heard its human cry.

The Hardest Metal
Platinum (Pt)

What draws prospectors to Earth's crust
in search of only a millionth
of one percent,
or in nickel
merely one out of every
two million radiant parts?

—A metal so hard,
to melt it: 1,775 degrees.
To boil it: 2,000 degrees more.
In open air
it will not tarnish
at any temperature.

In sixteenth-century Europe,
miners cursed the sandlike grains
grittying veins of copper and nickel,
or clouding chunks of cobalt
with flakes so fine
none dreamed
the tiny malformed octahedra or cubes
were lustrous, silver-blue
nuggets of strength.

The question, Italian poet-physician
Julius Caesar Scaliger said,
was how to smelt
what ancient Egyptians
had so elegantly fashioned
into necklaces, rings, and diadems.

Two hundred years later
with Spaniards exploring in Peru,

astronomer/author Antonio de Ulloa
stumbled across a new element
and chose to name this Incan silver,
unearthed at the River Pinto, *platina*
—our platinum, with other metals
hard-rock mined these days in Sudbury
or washed from Russian alluvial streams.

It will not react.
It will not dissolve.
It will not disclose itself
—why it finds its way
into a pen nib, a crucible,
a surgeon's precision instrument,
electrodes,
even optical glass,
or coating for missile cones.

In acid-making, refining, and engines,
above all other catalysts
the best is platinum,
even adorning a beautiful woman
sly and magnetic
—but resistant.

The Radioactive Metal
Uranium (U)

Ground down to a silver powder
what metal sets itself alight?
—Uranium.

Upon a darkened plate it glows
a ghostly tracery of bone
where X-rays burned,
or casts green-yellow shadows
from a shard of Naples glass
stained 2,000 years ago,
even delicately tints
flesh tones, sky, and leaves
in faded archive photographs.

Drawn toward magnetic fields,
it lets a gyro compass shiver
as if almost alive,
yet 1.6 times more dense than lead
its weight helps sink a spacecraft
back to Earth.

From isotope to isotope
it cannot stop or slow decay.
Its shrinking into lesser selves
radiates a chain of death
beyond a miner's burning lung
throughout kilometres of air and ground.

With thorium, uranium explodes
energy intense enough to boil
magma round Earth's centre.
Bombarded in a nuclear reactor
a mere lone tonne will fire

40 million kilowatt-hours,
as much as 20,000 tonnes of coal,
while breathing 90% less carbon
to our thinning ozone.

Uranium
—to warm, guide, and heal
or blow our glowing dust to hell.

Part III
Gem Lure

Oracle Stones

Listen to the stones'
eerie rattle as the tide rushes in
thousands clattering, grating, rocking
until waves pull back and suck down
pebbles slowing to shallow eddies
where water tinkles between.

What is it the stones would tell?
Choose one each to summon
earth, air, fire, and water,
plus a fifth, for asking:
pink, to loosen from silence
a whisper longing needs to hear,
a sparkling Trickster, to astonish,
or a holed Spirit stone,
to dare the truth.

To charge each pebble with meaning
pass it through incense and charcoal
smoke from essential oil.
Cast your fistful,
and the two pebbles that tumble
farthest and nearest the question stone
like forks on a divining rod
will quiver up an answer
out of the healing Earth.

For if, as a distant oracle chanted,
opal and other precious gems
are Earth's fancies in sleep,
surely tinted pebbles and shells
flinging from foam to sand
must be the sea's dreams.

Gem Lure

I

It has to do with light,
setting a gem to warm
in the sun on a window sill
or laying it out
two nights before a full moon
to stoke the glimmer of mystical powers.
For long ago, its glisten and sparkle
were deemed the efflux of stars,
a gem, the cosmos' reflection.

—But also Mother Earth's gift.
Bury it briefly, charge in salt water,
then rinse it clear in a spring or rain,
and a gem
to our simple eyes appears
not subatomic quarks' invisible whizzing
but flat, symmetrical facets
shining, solidified
as one.

Smooth, they calm
like the quietened space
where a gemstone chooses you
by cooling or tingling in your right palm,
drawing you ever closer, fonder,
through minuscule windows
to channel from deep within
and amplify subtle forces
that guide and ground
or from the top of the head surge down
the arm and out through the crystal
to centre, guard, and heal a friend.

II

But what is light? To science,
electromagnetic vibrations
in waves from 350
to 750 nanometres long
fanning out a "visible field"
into the rainbow of our spectrum.

Red, orange, yellow, blue, violet, green
vibrating as one, cast a white light
unless in the crystal's lattice
a fissure, gas bubble, or metallic
bits absorb particular wavelengths.
If all are swallowed, the jewel is black.

Long absorbing deepens a gemstone's hue,
and multiple mirroring facets beneath
sharpen its crystalline brilliance.
What of its outer reflecting? The lustre
can glow beyond waxy or silky
to bright as adamantine.

A feather crack billows through moonstone
a delicate opalescent light
or fires agate iridescent.
Feldspar glitters
from an included leaf
while a needle slits a sapphire
into a gleaming cat's eye
and opens silky canals
through a ruby.

—No wonder
we hold our breath
and stare into gem after gem.

Lithotherapy

Charged, the quartz worn on your wrist
vibrates at 30,000 cycles per second
humming out time
—physics' "piezoelectric effect".
Not only crystals, some say,
but we, also, vibrate
at frequencies a gifted few envision.
In the aura of our magnetic fields,
such energy is so subtle
the stresses of love and loss,
or a mere cold, or tippling
tilt it off balance.

The cure? Nine centuries ago
Abbess Hildegard von Bingen
described in her manuscript *Physica*
how gems had formed as God's blessing
from fire interacting with water.
Thus their radiance ever after
burned with elemental power.
Fondle a rounded stone
to let its curves soothe.
Grind it to a powder
for pressing into a pill
or mix as poultice paste.
Dissolved to an elixir,
ruby water cures stomach cramp,
a sip of diamond water illuminates.

Some healers meditate, or trust
to touch and lay a stone along
one of seven human chakras:
*   To ground emotions and recharge the body,
     upon the *Root* of the spine apply

red-streaked black hematite.
* Two fingers lower than the *Sacral* navel,
a crimson garnet clears and fires
passion, making women fertile.
* Rubbing amber's yellow resin
round the nerve-thick *Solar Plexus*
rouses courage and our vital essence.
* An emerald's green, if left to glow
upon a heavy *Heart's* throbbing,
blossoms love and well-being.
* Nestled in our *Throat's* hollow,
cool blue aquamarine inspires
canvas, symphony, or canto.
* Centred on the *Brow's* Third Eye,
from outer darkness, as within,
moonstone quickens intuition.
* To flash thoughts to an astral plane,
raise a diamond's ice-white spires
to the *Crown* chakra.

Enough on mystics.
What is healing's truth?
—Mere placebo, or gems' quantum physics?

Amethyst

Remember a cringing maiden
goddess Diana hardened to stone
when tipsy Bacchus loosed tigers upon her?

As if frightened all over again,
when left in the sunshine
an amethyst pales,
but flooding its facets with X-rays
recharges the gem's deep tones,
as Bacchus,
once sober and overflowing remorse,
poured a libation over his victim
turning her blue-red as wine.

Not so, according to science:
what tints the crystal
is the mineral iron.
Moreover, an amethyst's dark pointed tips
always root within geodes,
never on open ground.

For centuries, nonetheless,
respected lapidaries described
how amethysts had power to guard
not only a soldier or hunter
but any stumbling from passion,
contagion, or having too much imbibed.

Peer closer.
In an unheated gem,
that rare inclusion you spy,
is it perhaps
Bacchus' thumbprint,
or a tiger's black stripe?

Hematite

Sawed apart,
this black-grey stone
"bleeds" red iron oxide dust,
why Greeks called it *hema*: blood
that soaked from wounds into battlefields
and hardened wearers to become invincible.
Why, too, its mixed red ochre painted
cave walls, tombs, and beloved dead
for astral travel to the afterworld.

If ball or kidney shaped,
it charmed a surgeon's scalpel
against a hemorrhage or messy birth.
In egg-white paste, it shrank an eyelid tumour,
while a lump held to the forehead
drew a fever off.

Troubled? In a darkened room
gaze into a candle's wisdom
glimmering over hematite.
Blended from both earth and fire,
it counterbalances yin and yang
and grounds a dizzy spirit
through this simple mathematics:
to cancel negativity
add a plus: love.

Garnet

Red as pomegranate seeds
Greek god Hades gave Persephone
to lure his queen back to the underworld,
and petrified from goddess Isis' blood,
a garnet eased high Egyptian dead
from this incarnation to another.

So too, when 40 days and nights
flooded evil from the world
and Noah's kin
shivered in the ark,
to brace their hearts
and quiet bawling animals,
high above the prow he hung
a traveller's giant garnet, glowing
crimson hope against the leaden clouds.

Why, nestled close to diamonds, fiery
pyropes were a dragon's favourite
dug to bribe into its secrets.
A magic so potent
(iron's magnetism?)
forced back nightmares and the Evil Eye
and stirred a lower limb's circulation
to heal a snake or insect bite,
even battle injury.

A breastplate garnet charmed each Amazon
and nineteenth-century Indian rebels
shot the gems instead of bullets.
That hard, they have empowered ironmongers,
smiths, farmers—any who use metal—
but also forge deep bonds for lovers.
So energetic is their power

drops of goddess Isis' blood
are spattered everywhere on Earth.

Amber

Inside a greasy golden chunk
for 50 million years, a tiny bubble
poised between pine needle and a bug.
Rubbed, true amber still attracts
a hair or particle of dust
with a negative charge.

And so *elektron* was its Grecian name
when from grief three sisters turned
into poplars by the River Po
and once a year, shed resin tears
in memory of sky-driving Phaethon
too young to rein Sun's snorting stallions in.
Instead, the blazing chariot swerved
so low it scorched skins in North Africa
until Zeus zapped him with a thunderbolt.

His plunging fires, fishers claimed,
congealed beneath the sea and washed
as amber globs toward ancient Baltic shores.
Not so, said hunters. Urine from the lynx
or, in China, drops of dragon's blood
hardened to such yellow charms.

Wherever amber formed,
for balancing the body's chi
it was a pharmacopoeia of cures.
Clutch a lump to cool a fever.
A beaded choker shrinks a goitre.
Its liniment will quiet whooping cough.

Heated on a brick, its acid fumes
stop a runny nose and tonsillitis
and ease labour pains.

For a gentler fragrance,
powder some and set alight.
Amber's smoke is sweet as incense
above which—if you dare—
bare yourself and crouch
to cure a hemorrhoid.

Suspicious husband? Tuck it in
the bosom of a bedded wife
and she'll confess.
Yet though a potent sleeping drug
that dreams the dying to the afterlife,
no amber can withstand a perfume dab.

Emerald

In folklore, claimed colourless at birth,
an emerald would "ripen" green
in a gryphon's nest.
In truth, it bubbled up
through shale and schist,
altering cavities or plugging veins.

That's why
still so calming it is
to gaze into an emerald's *jardin*
and wonder how such dainty foliage
grew from crystal bits in fissures
magma greened with chrome.

Green not only soothes the probing eye
but blurs and softens memory
while tuning desire
to higher levels of knowing
—a revel in eternal spring
through its vibrant now—
the dream 4,000 years ago
as Cleopatra's mummies bore
emeralds to reach their afterworld.

Place the gem beneath the tongue
to block a spell, foretell,
or taste the truth.
If a lover strays, it splits.
A sharpener of wits and intuition,
it also heals with words made eloquent.

They fable how the Holy Grail
was sculpted from a single emerald
knocked from plunging Satan's crown

to form a chalice for the Last Supper,
and after, transubstantiate
wine into Christ's blood.

Moonstone

Advancing through the lunar month
across a moonstone's yellow skin
like a millet seed,
its bluish schiller swells to full,
then slowly thins again
into a crescent

—why mediaeval lapidaries claimed
this shifting gem had solidified
from lunar beams
and why its lustre was prescribed
to heal: the body, as it waxes,
the psyche, as it wanes.

Beneath a pillow,
moonstone deepens slumber,
but dream of it—there's danger
unless held in the mouth.
Then neither tears nor rain
will haunt the lone night wanderer.

At full moon, rolled beneath the tongue
the gem lets lovers taste
what will befall.
Worn while making love,
its light becomes an embryo,
and hung in trees, births abundant crops.

Others suck the gem to peel off madness
and glow into clairvoyance
through a Third Eye.
Some grind it to become invisible
or change dross into gold
through alchemy.

For deep inside, so Eastern seers believe,
there lives a goodly spirit
who will hypnotize
if you gaze into a swaying stone,
but scientists define as an inclusion
fringed with cracks so like a centipede,
its bluish opalescence
not magical at all, but "interference
phenomena" within its innermost layers.

Aquamarine

To mollify sea deities,
ancient lapidaries prescribed
blue amulets carved from aquamarine
whose inner lapping soothed
and as seasoned sailors believed
wore away the dark coast of worry.

Others cast their woes inside the gem,
then soaked it in a little bowl
beneath the waning moon.
Perhaps within a day or two
where crystals cooled and brittled
a six-rayed star would fan and twinkle
love light toward a long marriage,
restore youth, hope, and friends,
or calm a throbbing tooth.

Today Brazilian pegmatites
host the clearest and the bluest,
named (her birthstone) Santa Marias.
A famous one, unearthed in 1910,
was heftier than a bongo drum:
110.5 kilograms
—an amulet with cleansing tears enough
for a thousand sailors
not to drown.

Diamond

From 150 kilometres under the Earth
molten carbon funnelled up
as carrot-shaped pipes.
Erupted through a volcanic cone,
over eons, its kimberlite weathered
from igneous blue into yellow ground
where diamonds may be strewn
formed so deep, so pressured,
of all crystals, the hardest.

No, they never burst to light
from thunder bolting into rock,
as India's royal legends claimed,
nor, when moistened with May dew
and left three nights in darkness,
will two diamonds reproduce.
Neither can they multiply good fortune,
honours, strength, or youth,
even if fortified with iron.

A diamond in a gold setting
braceletted upon your left arm
drives away no midnight spectres.
Its icy sparkle can't defy the Devil
or dim to show who's guilty of a crime.
It never sweats a warning: there is poison.

Although in myth this "King of Gems"
flashed the ancient sun's perfection,
today allied with Mars and men
and mined by open pit or shaft
its rough transparency must serve
for abrasives, drillers' bits—and love.

The Magic of Stones

Stroke a flawless rock,
and you become
—invisible!
What if another
gleams in the dark?
Or stings your tongue?

So many shapes tickle your palm,
be they needles and threads,
opalized animal bones,
stiffened straight as a pencil, a cross,
or opening to a rose,
a chrysanthemum.

Sniff pine,
and you may hold amber
from soil perspiring under a sinking sun.
Rubbed, it becomes
electric to chaff and linden,
like lodestone pointing north from south
that once, it's told, in ancient Magnesia
locked a shepherd to the ground
by his iron-nailed soles.

Or look above, to the stars
for fossils falling to seed the earth,
and meteoritic iron to forge gods' armour
strong as the quarried mountain slope
healed by growing marble
to fill its wounds.

Though plunged six feet into the earth,
a thunder egg may guard against lightning,
each rumble drawing it one foot nearer the air.

Calabrian peasants claim that when
a blue thread holding it over a fire
does not burn,
the egg is one more talisman
to guard against snakebite,
spark courage, or love.

But which,
tight in your fist,
will make you invisible?

Clay

What gemstone?
We consider clay mere dirt,
but ancient deities knew better.

In Africa, the Earthcreator
laid down a "thought-design"
to cookie-cut the prototype of man,
while scoops from China's yellow clay
shaped no peasants (they were mud),
only scant aristocrats.

In Babylon, like flatbreads, unleavened,
clay was pressed with blood
and a burnt god's flesh.
In Egypt, it was mixed with straw,
whirled on a potter's wheel.
Adam's rib made woman,
while the Moslem recipe included
dust, breath, a clot, tissue,
and a drop of semen.

But given our multi-tinted skins,
why not sculpt us rose quartz,
amber, or obsidian?
Only clay binds like a magnet.
Its nitrogen and carbon ions,
being negative, attract
within the soil enough oxygen,
phosphorous, and hydrogen
for a biological reaction.

Simply, clay shifts back and forth
from moist, to dry, to moist.
Thus, observed the Koran,

organic molecules can mesh long chains
to form deoxyribonucleic acid,
our human blueprint (DNA).

Part IV
Going Under

Prospector
(for Eira Thomas)*

Who bush-planes into the Barrens
300 kilometres northeast of Yellowknife
where even in sunshine the mercury plummets
to 40 degrees below,
skin freezes in 30 seconds,
and snow whites out all traces to navigate by?

Who wants to trek up the winter road
to hunker ashiver in wind-yanked tents,
where eating means reaching outside the flap
to hacksaw another steak from the ice,
or in summer, to toil for a trout
by casting along a shore
where grizzlies and wolves prowl, sniffing?
—All this, to search leftover volcanics
for crumbly, dark kimberlite.

Picture a blue-eyed geologist,
tall and slender, who at a mere 24
one May morning, strides over the hills
to the edge of a shallow stream,
slings the pack off her back,
and with a rock hammer
picks a chunk loose from the bed
in search of that mineral signature
magnetics had blipped to a distant screen.
She holds up and squints at a shard
dark and unevenly grained.
She *believes*.

—Enough to keep charting
for month after month
a course farther west

83

and by spring, to walk upon water
still frozen at Lac de Gras,
to fly in a crew
and round the clock drill into ice,
pulling up core lengths into the shack
where one that she crushes with her rock hammer,
sloshes with water, and floats the lighter bits off
reveals in the "heavies" on bottom
pyrope deep-purple garnets,
the hint that diamonds are hidden,
as later a lab in Toronto reports
—20 micros, in fact.

Day after day, as the temperature rises,
along the lake edge, ice loosens,
tinkling thin vertical shards.
The surface softens and slushes.
Groaning, it sags under tonnes
when she flies in a wider drill.
Over ice where the bit rams through,
black smoke billows up.
Water begins to pool.
It rises above her ankles.
It slaps up the walls of the shack.
It floods past the drill crew's knees.

That flash along shore—open waves?
Soon fog settles in and thickens.
—What helicopter can land?
Still she keeps pushing the roughnecks:
Again! Ram the drill bit down.
A second target, a third
roars through ice,
through granite, through nerves,
through minutes fast melting away.

That night, she changes
the locks to the core-sample tent,
unplugs computers, and cuts the telephone line.
Under her pillow,
pried from a kimberlite length,
grinning, she stashes a 2-carat stunner,
mark of the highest grade cluster
of diamond pipes in the world:
138 million carats.

All because she *believed*.
For decades, the majors had scoffed,
"Mine diamonds in Canada?—Never!"
until Chuck Fipke, a dogged geologist,
unearthed a studded kimberlite pipe
in the Barrens' Archean rock.
Yet even after the staking rush started,
who listened to a young woman,
or risked their own bottom line
to drill that *one extra* hole
and gamble that under thin ice
the biggest of mining dreams sparkled.

* Eira Thomas is now in the Canadian Mining Hall of Fame. Her
discovery of pipe A154S eventually became part of the Diavik
Mine, owned by Canada's Aber Diamond Corporation and
London-based Rio Tinto plc. Chuck Fipke's discovery led to the
Ekati Diamond Mine, owned by BHP Billiton Limited.

The Bottom of the Sluice-Box
(a found poem*)

So long as you live, you shall not forget
the shimmering water glancing over
bits of snow-white quartz,
green-stone, and jasper, over glinting mica
and crystals of garnet, bringing out
the keen colour of all
the polished and far-travelled pebbles,
the red of the "ruby" sand, and the heavy
magnetic iron, coal-black and sunk to bottom,
where it lies
striking sharp contrast
against the yellow gold.
To take "the stuff" up
so, in your hands,
dripping
and shining and mixed with the elements
is to get the fine flavour of the richness
of the King of metals.

* These words, reset as lines in a poem, are taken from the article
"Pleasure Mining" by Elizabeth Robins (1862-1952), *Fortnightly
Review*, March 1, 1902, http://www.jsu.edu/depart/english/robins/
alask/pleasmin.htm, p. 486.

Underground

Suspended 1,000 metres into the mine,
through wire mesh under your boots,
you glimpse lights twinkling
a further 500 metres down,
a black sea netted in stars.
This is the underworld.

Here, as time descends,
frozen rock thaws, leaning
inward warmer and warmer,
and manmade mists must billow away
invisible crystalline silica dust
breathing out deadly scars.

Snared in the wire cage,
pinched between forklift and loader,
a few shivers, a lurch—you're dropping.
Vertical rubber wheels hurtle you down
blurring level on level
as Earth's heat rises
and rubberized coveralls
soften and glisten—at more
than 35 degrees Celsius now.

A yank—
a bungee-cord bounce—
the cage whiplashes and stops.
Holding your ears
you edge out through thunder
and blink from the eye-stinging dust
into a cavern 35 metres high,
hub of a hollowed-out vast wheel,
its spokes angling off into distant drifts
where under low bolted ceilings

shaking helmets rattle pneumatic drills,
while trucks rumble rocks back and forth
loading more clattering chunks to roar
down billowing metal chutes beneath
to huge metal jaws crushing the ore.

Deep along one drift,
a manway is the only hole out
to another shadowy level below.
You ease in hips, then shoulders, begin
the long and too steep climb down.
Rung by rung you grip harder.
Wedged in so thin
—how can you breathe?—
what if you freeze halfway?

At bottom,
splashing through seepage,
thank God, you are not alone
as helmet lamps are clicked off,
drowning all in a darkness so black
you can't even focus, let alone guess
left side, or right, or up.
Is this how it is on death's rim
—senses dissolving into oblivion?

An hour after,
uncaged above ground,
you blink into midday sun.
Shoulders unclenched at last, you shake
the lead geologist's steadying hand:
Thanks, for a tour of the mine.
Indeed, it was quite an adventure.
Great story to entertain friends.
—*But you'll never do it again.*

Working the Mines

1. Artisanal Miners

To decorate his skin for sacred rites,
an early Stone Age man wandered
searching for hunks of red ochre.
To dig with, on a notched stick he bound
a cobblestone or knapped flint
or, sharper, obsidian.

One day, wading along a stream
he picked up a stone that glittered
and, oh—how it dazzled the women.
Wide-eyed for more, others learned
to tilt a basin, slosh around gravel,
and pluck the brightest chunks.

Long after the visible nuggets were culled,
scavengers rinsed sand through a fleece
they laid to sparkle dry in the sun,
then spread the wool over a fire.
Glittering grains fused into lumps
easily picked from the ash as it cooled.
In time, quicksilver drops were found
to stick to and lure out the gold
—how metallurgy began.

2. Slaves

Millennia later, underground
lighting a fire against a rock face
by morning cracked off a slab of ore.
Wherever the grain stained bluish-green
hinted copper might lurk within
or lead, perhaps, if brown.

89

Coughing through veils of mercurous smoke,
with only bone, antler, and hardwood
slaves sweated to pick, lever,
and hew out block after block,
then hammered each into chunks
and pounded the ore to a fine powder.
To burn impurities off
they stirred in barley, salt, and lead
and in a clay cupel fired the mixture five days.

Three torturous years spitting poison and blood
crumpled and broke the sturdiest slave,
but any still straining to sink a shaft
by 20 metres down sobbed thanks:
their lethal chiselling had to end
as groundwater flooded in.

3. Shareholders

By five centuries B.C.,
mining the parched steep hills
where Laurion overlooked the Aegean,
every Athenian boasted one silver share,
160 million ounces together
$10 billion worth.

Now as many as 2,000 shafts
barely 1 x 2 metres across
sank to 100 metres deep.
Lit by olive-oil lamps,
slaves inhaled fresher air
drawn down a long system of vents
between well-spaced rock pillars
left standing to support drift roofs
and never chipped into—the penalty, death.

Bodies glistened, hacking out ore
and dragging it up wooden ladders
or hoisting sacs to the surface on ropes.
Above, cement-ridged tables
washed the hammer-crushed ore,
recycling the water in cisterns 10 metres wide.

A small bellows furnace fired out metal.
A clay cupel melted off lead to slag,
smelting silver 98% pure
and plentiful enough for minting
the first silver coin, a drachma,
that financed 130 ships
for Athens to conquer Persia and the Aegean
till Sparta's fortress at Deklea
strangled the silver supply.

4. Free Miners

Heaped a-gleam with silver and gems
fairy tale caverns clanging with dwarves
were modelled on Saxony miners—free men—
and many did pile up wealth,
when Markgraf Otto of Meissen
(who himself became Otto the Rich
and fathered a line of Europe's kings)
let any stake claims on public land
by paying a royalty of 10%.

Skilled labour in 1000 A.D.
did better than hoist groundwater
from shafts in leather buckets by hand.
For Otto the Great
at the king's Rammelsberg mine,
30 cramped years, undaunted, they carved
with only chisel and hammer

a drainage tunnel 900 metres long
to splash out the Harz Mountain side.

5. Technocrats

Glittering on a palm
or locked away in darkness,
a chunk of gold delighted, but little else
—why the Dark Ages lasted 500 years:
Barbarian raiders had hoarded their spoils;
Crusaders squandered sacked gold in the east.

But once the metal was ground and melted,
rounded and flattened into small coins,
the pieces were light enough
to carry far distances,
to count out into a hand,
to clutch through fire and flood.

Why pay with feudal acres or crops,
when coins hired soldiers full-time
not only to defend a farm
but as in Alexander's era
to march from Macedon to Asia,
and Caesar, to Egypt and Gaul from Rome.
Victory clinked from a bagful of coins,
and battles and kingdoms were lost
because of an emptied purse.

Coins a peasant could save to buy land,
clear it, cottage it, and from the rent
yield the shire more taxes,
creating jobs: bookkeepers,
clerks to sort and file records,
administrators to oversee and plan.

As deal-makers sought out each other,
many great trading cities blossomed:
Paris, London, Venice, Florence.
Beyond a castle or cathedral
grids of civic palaces, ports, and canals
spread and flowered with artists and craftsmen.

To pay for a seaport, a courthouse, a road,
instead of a wagon sack-full with coins
the first letters of credit were scribed
backed by great stacks of bullion
housed in a new guarded coffer
named for its walls, a *bank*
which funded, in turn, their local exports
—Venetian glassworks, Umbrian looms—
righting the balance of trade.
Wealthy citizens flourished as patrons
acquiring Classical learning and art
—financing the Renaissance.

6. Engineers

As centuries passed, technology automated
from tunnel pails, to dams and sluices
that powered large water pumps.
Sleds and wheelbarrows rolled out the ore
to ropes where a human treadmill above
improved to a windlass trod by horse.
Mining tightened its iron fist
firing a country's cannon and trade
to wealth that homespun an industrial age.

How far mining has come,
from a Stone Age cobblestoned stick
to the drill rig's 40-kilogram tricone bit,
its bullet-shaped tungsten carbide teeth

ripping straight down through rock
100 rotations per minute;
how far, from bonfires lit overnight
to explosives remotely controlled
blasting whole walls of ore,
too massive for piling into slave's baskets
but not for a 10-tonne Load-Haul-Dump
12-metre-long steel mucker.

Gold is explored no longer solely
by a lone man scouring a stream
and tilting a simple pan,
nor more subtly by Geiger counter
or a bush plane low overhead
swinging a magnetometer aft.

Even farther above,*
a satellite now probes Earth,
imaging hectares of lonely terrain
and beeping data to a computer
to e-mail prospectors on the ground
co-ordinates where to mallet in stakes.

Far below,
in near darkness,
through GPS-gizmos on shovels and dozers
to surface computers
a satellite diagrams every move,
every hazard throughout the mine,
and fixes precision crosshairs
on where next to trigger
a blast in a seam.
Even the drill bit houses a delicate sensor
tracking each rock, stratum, and ore
it chews a hole through.

Once every drift is emptied,
and all the miners have gone,
a satellite plans and scans through time
the angles, cuts, cleaning, and seeding
for hills, valleys, wetlands, and bush
reclaimed, again to be green.

    * The Mining Automation Program in Canada

Horror Stories

1. Iron Mountain*

Down mine ceilings and walls
trickled electric blue, purple, and green
10,000 times more biting than battery acid.
Lying overnight in their puddle,
a steel shovel once dissolved
and ripples caught on fire.

Yet in their blistering rainbows flourished
a life form none before had seen:
an acid-eating microbe.

Brewed out of heavy metals
oozing yellow-boy toxins,
it was the vilest of water
that over many kilometres leached
along tributaries and fouled
a city's drinking supply.

> * The Richmond Mine in the Klamath Mountains of Northern
> California, near a tributary of the Sacramento River headwaters,
> 9 miles north of Redding, Shasta county, was neglected from the
> 1860s to 1963 and will continue leaking for 3,000 years.

2. Cyanide

In Berlin, 1704,
heating and mixing dried blood,
two scientists chanced upon "Prussian Blue",
a dye so potent
a rice-grain worth,
if swallowed, could finish a man.

Now white, crystalline sodium cyanide
is stirred into water and sprayed
over heaps of crushed ore
where after dissolving out the gold
it gurgles down to pit bottom,
eventually to be drained.

What if once more it leaks astray
loosed by warming in sunshine
or trickles under mine tailings
through groundwater into streams,
poisoning fish by the tonne
and fauna by thousands?

3. Hole*

Picture a shovel massive enough
to scoop up two SUVs to dump
into a truck two storeys high.
It sliced off a mountain top
and tier by tier, spiralled a hole
so deep it could swallow the CN Tower,
so wide its diameter awed
visible at 200 kilometres distant
—even to Space Station astronauts.

For gouging minerals from the earth,
the pit seemed a cost-effective plan
for years—while being worked,
but once the walls had been scooped clean,
bulldozers gone, the pump cranked off,
groundwater leaked back in.

Out floated heavy metals and sulphur
that rose to an orange lake so toxic
it slaughtered alighting birds

and like a backed-up drain
sucked the last clean dribbles away
from already parched creeks and farms.

    * The mile-wide Berkeley Pit in Butte, Montana

4. Silicosis

Sniff in a silica crystal,
lung tissue threads around it
while scarring tightens the knot.
The bigger the nodule grows,
the harder it becomes to breathe,
the frailer the body against bacilli.

In *De Re Metallica*, 1556,
Agricola wrote of fevered stonecutters
coughing with chest pains and turning blue:
"Miner's Asthma", "Potter's Rot", or "Phthisis"
from years underground or in a pit
gulping the dust-choked air.
Now, 450 years after?
A high-pressure power blaster
swirls more—and finer—everywhere.

5. Prophecy

South of the Arctic Circle,
alongshore Great Bear Lake
down the "Highway of Atoms"
for 2,100 kilometres
Dene men* lugged "the money rock",
uranium ore, in coarse 45-kilogram sacks.

They slept on its warmth
on barges, portages, and docks,

thickening their lungs with its dust
12 hours a day, 6 days a week,
4 months, year after year,
to fuel a nuclear bomb.

At home, they ate fish from the tailings water.
Children played blocks with the grimy ore.
The tainted sacks wives sewed into tents,
while husbands away on hunting trips
camped for days in a wilderness
alive with radiation.

Heads ached, hair thinned,
hearts palpitated, while bleeding
pooled as splotches beneath their skin.
And what of the offspring to come?
How many more oddly aborted
or grew into lopsided lives?

Too soon, unseen, the mines' radon gas
or fallout swallowed and breathed
coughed each man to death,
and as a grandfather once had prophesied
Great Bear Lake curled yellow and foul
as uranium widows' grief.

    * Seasonal employees of Eldorado Mining and Refining, which
    was owned until 1988 by the Government of Canada

Reclaiming the Earth

Gliding along the hills
who would be able to tell
metal was ever mined here?

For greening to begin,
haul away chemical drums,
all diesel and gasoline tanks,
and bury leftover pipeline lengths
under waste rock and tailings.
Unstring the hydro line.

Where an ore stockpile rose,
furrow a metre-deep clean swath
and shower over the heap-leach pad
bacteria that gobble up cyanide.
Deliver the pregnant ponds
by pulling out their liners
and fill the depressions
with boulders and pebbles
free from toxins that might dissolve.

Five metres from the edge,
hem in the open pit for flooding
within 100 metres of the top ledge
and blast down the uppermost 2%
to brace and block in
lower acidic walls.

Whatever structure cannot be moved, dismantle.
Whatever cannot be recycled, burn.
Whatever cannot be burned, bury.
Wherever foundations remain,
fracture and crumble
the concrete slabs.

Angled for water to drain,
smooth crushed stone into slopes
as if the landscape had never eroded.
According to government regulations,
"Cover with 0.3 metres of topsoil,
recontour, and revegetate."

To close off a sleeping world,
at every entry pile soil and logs
matted with local quickening roots.
Scar incoming roads,
and soon networks of fissures
will sliver up grasses and weed.

For 3 years more, the local steward
must wander, watch over the site,
and monitor surface water,
and every 6 months,
an inspector fly in to gauge
how far the land has regrown.

Years after,
who will be able to tell
metal was ever mined here?

No one
riding the wind,
not even the birds.

Pest Control

Over millennia, like mites,
thousands have burrowed
under Earth's skin.

Stung, half waking, a stratum twitches,
and deep within her seismic gloom
another rockwall cracks.

Sometimes, coal dust swirls through tunnels,
tickling, and long-held breath relaxes
into her catastrophic sneeze.

Let gas fill opened pockets and smother.
If not, her rumbling sets it afire
to blister away the nuisances.

One day, she looses scalding steam,
another, floods groundwater up,
or pulls in her gut, tight.

Crush. Drown. Burn.
Blow up. Suffocate. Gone!
She settles—just for an eon—back to sleep.

Epilogue

Petrologist

Imagine deep in the Earth
crystal's ions intermingling
to bristle geodes with amethyst,
or sediments pressed under heat and time
meeting magma's massive mineral folds
where the purest grade ore
glimmers Plutonian secrets:
rifts and strata forming, reforming,
by millimetres moving whole continents.

Craving such jagged edges, he bends
to maps and data far into the night
by the light of a sharp desire
uncertain if
he can fathom the steep
darkness of the way down.
Between the compass points of Death and Love,
whorls the centre's magnetism.
His instrument is intuition.
It watches the numbers dance:
There, under there,
start to dig.

Blunt at first, a pneumatic drill,
deeper, a hammer pick
risk the wrong fissure,
force the layers apart,
and he goes on believing despite
one tap askew that shatters the opal.

If time drags slow as tectonic plates
and patience smokes like a match,
does it matter,
striking bornite or sulphur

and not aquamarine?
Does it matter
even that he digs at all?
No more than that rain falls
and hydrothermal springs steam.

In sunlight, one by one,
he lays out his stones,
loving their tints,
silky or ragged faces.
He fondles, speaks to them,
wonders when they will whisper back
where each formed,
what it grated against,
where some day it would go.

Even gneiss heavy
and long pressed in,
an ancient and beautiful few aspire,
like restless swallows, to spiral through air
to taste sky fire and briefly transcend
lava's relentless downward pull.

And so he raises
their heft and lustre
high toward wind and sun
and hopes
at some far edge
hands uplifted await
their shimmer and angles,
his humble translation
of eons into words.

About the Author

Susan Ioannou of Toronto first became interested in geology as a theme while her son was completing a PhD. Exploring the science of rocks and minerals from a poet's perspective was a fascinating and refreshing change from writing personal lyrics. Ioannou's fiction, articles, and poetry have appeared across Canada. Winner of the 1997 Okanagan Short Story Award and twice a finalist in the CBC Literary Awards, in 2002/2003 she received an Ontario Arts Council Works in Progress grant to complete *Looking Through Stone*. Her other books include *Clarity Between Clouds* (Goose Lane Editions), *Where the Light Waits* (Ekstasis Editions), and *Coming Home* (Leaf Press), the children's novel *A Real Farm Girl* (Hodgepog Books), and the handbook *A Magical Clockwork: The Art of Writing the Poem* (Wordwrights Canada). She is a member of the League of Canadian Poets, and a past Associate Editor of *Cross-Canada Writers' Quarterly/ Magazine*.

MARQUIS

MEMBER OF SCABRINI GROUP

Québec, Canada
2007